WOUND CARE HANDBOOK

I0474616

Published by

Sri Lakshmi Services

© Vasu K Brown, MD

Published by

Sri Lakshmi Services

© 2009 Vasu K Brown, MD

Introduction

Wound healing is a complex and dynamic process with the wound environment changing with the changing health status of the individual.

The knowledge of the physiology of the normal wound healing trajectory through the phases of hemostasis, inflammation, granulation and maturation provides a framework for an understanding of the basic principles of wound healing.

Through this understanding the health care professional can develop the skills required to care for a wound and the body can be assisted in the complex task of tissue repair.

A chronic wound should prompt the health care professional to begin a search for unresolved underlying causes. Healing a chronic wound requires care that is patient centered, holistic, interdisciplinary, cost effective and evidence based.

Why Do Wounds Happen?

In any natural disaster the damaging forces must be identified and stopped before repair work can begin. So too in wound care the basic underlying causes and factors that affect healing must be identified and controlled as best we can before wound healing will begin.

Following are some of the common underlying causes or
factors, which may interfere with wound healing:

Trauma (initial or repetitive)
Scalds and burns both physical and chemical
Animal bites or insect stings
Pressure
Vascular compromise, arterial, venous or mixed
Immunodeficiency
Malignancy
Connective tissue disorders
Metabolic disease, including diabetes
Nutritional deficiencies
Psychosocial disorders
Adverse effects of medications

In many cases the underlying causes and factors interfering with wound healing may be mutlifactorial.

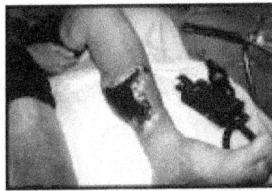

Figure 1 illustrates an elderly patient who suffered trauma when she banged her leg on a coffee table. She is on coumadin which contributed to the injury becoming a large black hematoma of old blood.

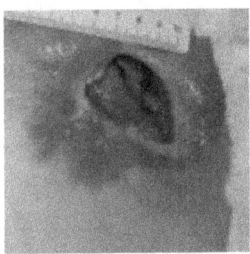

In figure 2 we see a young spinal cord injured patient with a chronic pressure ulcer surrounded by erythema.

Erythema could be caused by infection, irritation of wound fluid, incontinence or continual pressure to the area.

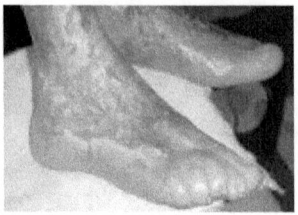

In figure 3 we see chronic ulcers in a frail elderly woman that has lower leg edema related to decreased mobility. The ulcer drains copious amounts of chronic wound drainage causing irritation to the surrounding skin. The patient sits most of the day in a dependent position which worsens the leg edema.

The clinician working in wound care needs to be a good detective and needs to consider all possible factors influencing healing.

How Do Wounds Heal?

Research work on acute wounds in an animal model shows that wounds heal in four phases. It is believed that chronic wounds must also go through the same basic phases1.

Some authors combine the first two phases.

The phases of wound healing are:
Hemostasis
Inflammation
Proliferation or Granulation
Remodeling or Maturation

Kane's analogy to the repair of a damaged house provides a wonderful framework to explore the basic physiology of wound repair

Hemostasis:

Hemostasis - Immediate Platelets Capping off conduits

Once the source of damage to a house has been removed and before work can start, utility workers must come in and cap damaged gas or water lines. So too in wound healing damaged blood vessels must be sealed. In wound healing the *platelet* is the cell which acts as the utility worker sealing off the damaged blood vessels.

The blood vessels themselves constrict in response to injury but this spasm ultimately relaxes. The platelets secrete vasoconstrictive substances to aid in this process but their prime role is to form a stable clot sealing the damaged vessel.

Under the influence of ADP (adenosine diphosphate) leaking from damaged tissues the platelets aggregate and adhere to the exposed collagen3. They also secrete factors which interact with and stimulate the intrinsic clotting cascade through the production of *thrombin*, which in turn initiates the formation of *fibrin* from *fibrinogen*.

The fibrin mesh strengthens the platelet aggregate into a stable hemostatic plug. Finally platelets also secrete cytokines such as *platelet-derived growth factor* (PDGF), which is recognized as one of the first factors secreted in initiating subsequent steps. Hemostasis occurs within minutes of the initial injury unless there are underlying clotting disorders.

Inflammation Phase:

Inflammation - Day 1 - 4
Neutrophils Unskilled laborers to clean up the site

Clinically inflammation, the second stage of wound healing presents as erythema, swelling and warmth often associated with pain, the classic "rubor et tumor cum calore et dolore".

This stage usually lasts up to 4 days post injury. In the wound healing analogy the first job to be done once the utilities are capped is to clean up the debris. This is a job for non-skilled laborers.

These non-skilled laborers in a wound are the *neutrophils or PMN's (polymorphonucleocytes).* The inflammatory response causes the blood vessels to become leaky releasing plasma and PMN's into the surrounding tissue4.

The neutrophils phagocytize debris and microorganisms and provide the first line of defense against infection. They are aided by local *mast cells*. As fibrin is broken down as part of this clean-up the degradation products attract the next cell involved.

The task of rebuilding a house is complex and requires someone to direct this activity or a contractor.

The cell which acts as "contractor" in wound healing is the *macrophage*. Macrophages are able to phagocytize bacteria and provide a second line of defense.

They also secrete a variety of chemotactic and growth factors such as *fibroblast growth factor (FGF), epidermal growth factor (EGF), transforming growth factor beta (TGF-___ and interleukin-1 (IL-1)* which appears to direct the next stage5.

Proliferative Phase (Proliferation, Granulation and Contraction):

Day 4 - 21 Macrophages, Lymphocytes, Angiocytes, Neurocytes,, Fibroblasts, Keratinocytes, Supervisor Cell, Specific laborers at the site: Plumber, Electrician, Framers, Roofers and Siders

The granulation stage starts approximately four days after wounding and usually lasts until day 21 in acute wounds depending on the size of the wound.

It is characterized clinically by the presence of pebbled red tissue in the wound base and involves replacement of dermal tissues and sometimes subdermal tissues in deeper wounds as well as contraction of the wound.

In the wound healing analogy once the site has been cleared of debris, under the direction of the contractor, the framers move in to build the framework of the new house. Sub-contractors can now install new plumbing and wiring on the framework and siders and roofers can finish the exterior of the house.

The "framer" cells are the *fibroblasts* which secrete the collagen framework on which further dermal regeneration occurs. Specialized fibroblasts are responsible for wound contraction.

The "plumber" cells are the *pericytes* which regenerate the outer layers of capillaries and the *endothelial cells* which produce the lining. This process is called *angiogenesis.*

The "roofer" and "sider" cells are the *keratinocytes* which are responsible for *epithelialization*. In the final stage of epithelializtion, contracture occurs as the keratinocytes differentiate to form the protective outer layer or stratum corneum.

Remodeling or Maturation Phase:

Remodeling - Day 21 – 2 yrs Fibrocytes Remodelers

Once the basic structure of the house is completed interior finishing may begin. So too in wound repair the healing process involves remodeling the dermal tissues to produce greater tensile strength.

The principle cell involved in this process is the *fibroblast.* Remodeling can take up to 2 years after wounding and explains why apparently healed wounds can break down so dramatically and quickly if attention is not paid to the initial causative factors.

When Does a Wound Become Chronic?

In healthy individuals with no underlying factors an acute wound should heal within three weeks with remodeling occurring over the next year or so. If a wound does not follow the normal trajectory it may become stuck in one of the stages and the wound becomes chronic.

Chronic wounds are thus defined as wounds, which have "failed to proceed through an orderly and timely process to produce anatomic and functional integrity, or proceeded through the repair process without establishing a sustained anatomic and functional result."

Once a wound is considered chronic it should trigger the wound care clinician to search for underlying causes, which may not have been addressed. Better yet, an understanding of the causative factors should lead us to be proactive in addressing these factors in at risk populations so that chronic wounds are prevented.

So Chronic wounds are stuck in either prolonged inflammatory stage or proliferative stage.

Basic Principles of Wound Care

There are three basic principles which underlie wound healing.
1. Identify and control as best as possible the underlying causes.
2. Support patient centered concerns
3. Optimize local wound care.

Optimize Local Wound Care

In 1962 George Winter described improved wound healing under moist conditions7. Despite that seminal work it is only in the last decade that the advantages of moist interactive wound healing have become more widely recognized and applied in clinical practice. Some of the advantages include the following:

Decreased dehydration and cell death. As described earlier, the task of wound repair requires the activity of a host of cells from neutrophils and macrophages to fibroblasts and pericytes. These cells cannot function in a dry environment.

Increased angiogenesis. Not only do the cells required for angiogenesis require a moist environment but also angiogenesis occurs towards regions of low oxygen tension such that occlusive dressings may act as a stimulus in the process.

Enhanced autolytic debridement. By maintaining a moist environment neutrophil cell life is enhanced and proteolytic enzymes are carried to the wound bed allowing for painless debridement. Further as discussed earlier these fibrin degradation products are a factor in stimulating macrophages to release growth factors into the wound bed.

Increased re-epithelialization. In larger, deeper wounds epidermal cells must spread over the wound surface from the edges. They must have a supply of blood and nutrients. Dry crusted wounds reduced this supply and provide a barrier to migration thus slowing rates of epithelialization.

Bacterial barrier and decreased infection rates. Occlusive dressings with good edge seals can provide a barrier to migration of microorganisms into the wound. Bacteria have been shown to pass through 64 layers of moist gauze. Wounds covered with occlusive dressings have been shown to have lower rates of infection than those with conventional gauze dressings.

Decreased pain. It is believed that the moist wound bed insulates and protects the nerve endings thereby reducing pain. Furthermore occlusive dressings often require fewer dressing changes, which may be uncomfortable for patients.

Decreased costs. While occlusive dressings have a higher per unit cost than conventional gauze, the reduced frequency of dressing changes and increased healing rates may proved to be cost effective in the long term.

While moist wound healing has clear advantages, debate continues on how moist is moist. Dressings should retain enough moisture to stimulate good healing and yet should not cause maceration or irritation to the surrounding tissues.

The Ideal Dressing

So how do we provide for good moist interactive wound healing? In 1979 Turner described the ideal dressing as having the following characteristics:

Removes excess exudate and toxins
High humidity at the dressing wound interface
Allows for gaseous exchange
Provides thermal insulation
Protects against secondary infection
Free from particulate and toxic components
No trauma with removal

Over the past 15 years an ever-expanding list of dressing products has come onto the market in an attempt to meet these conditions. Among these are the transparent film dressings, hydrogels, hydrophilic foams, alginates, hydrocolloids and the new antibacterials and biologic dressings or devices. There is however no magic "one-size-fits-all" dressing.

The clinician needs to become familiar with the characteristics of the different classes of dressings and to tailor the dressing used to the phase of healing, characteristics of the wound, the needs (and risk factors) of the patient and the availability and skill of the caregiver.

Wound Assessment dictates various treatment options:
Assessment of the wound is a prerequisite to the selection of an appropriate dressing.

1. Cause – Determine Etiology
2. Local Wound Characteristics
   a. Location
   b. Size – length x width x depth
   c. Wound bed – black, yellow, red, pink, undetermined
   d. Exudate – copious, moderate, mild, none
   e. Wound edge – callus and scale, maceration, erythema, edema
   f. Odor – absent, present
3. Patient Concerns: pain, persistant or temporary
4. Condition of surrounding skin – normal, edema, warmth, eythema
5. Clinical signs of critical colonization/local infection and infection

Dressings

1. Determine Etiology:

Redness and Excoriation of Skin

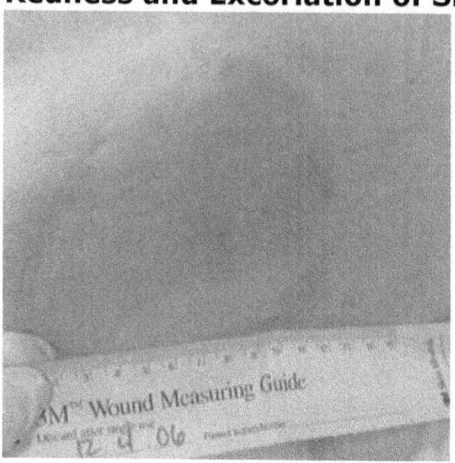

At the sign of excoriation assess for causes including fecal, urinary incontinence, any underlying disease conditions including Diabetes mellitus, Renal failure, NSAIDs usage, Coumadin therapy.

For first line of treatment use
Wound honey with A&D to wound bed cover with either small piece of unna boot or alginate and use either Mepilex border or Allvyn Gentle border dsg and change q 7 days. In the presence of excessive drainage change q 72 hours.

If wound honey is not in formulary use Triad cream with A&D. Triad cream application does not need a covering dressing.

You can also use Exuderm, Tegaderm with foam dressing for the above areas.

If the redness persists with treatment might be a sign of fungal infection use antifungal cream, Calmoseptine with 1% hydrotisone cream bid and prn soilage.

Open areas with excoriation:

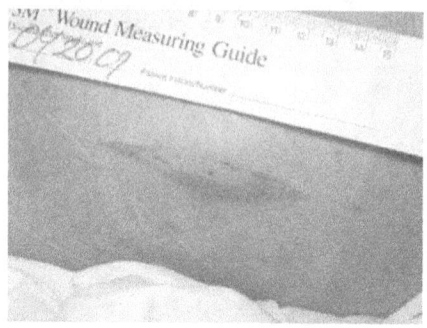

You can still use Triad cream with A&D with Unna boot strips on top.

Wound honey with A&D with Mepilex border or Allvyn gentle border dressing and change dressing every 72 hours if dressing stays dry or prn soilage change the dressing.

My personal choice – use wound honey with A&D cover with unna boot strips as 4x4 and a border dsg change daily.

General Guidelines for Wound Management

Wounds are divided into 4 Major Category

Wounds that are **Dry**
Wounds that are **Wet** – (**3 Subgroups**)
Wounds that are Wet are subdivided into 3 subcategory

Exudates – **Mild**
Exudates – **Moderate**
Exudates – **Large**

All wounds again has 2 varieties 75:25 ratio of Granulation and Slough
ie 75%Granulation with 25%Slough or 75%Slough with 25%Granulation

So the treatment is based on the way the wound category is for that week

If Dry with >= 75% Granulation use
1st choice Wound honey with A&D cover with Mepilex border or Duoderm

2nd choice Wound gel with Alginate with Mepilex border or Duoderm

If Dry with >= 75% Slough use
1st choice Hypergel with alginate cover with Mepilex border or border dsg change q 72hr

2nd choice Triad cream with A&D cover with Mepilex border or border dsg change q 72hr

Wounds with Exudates'
For **Mild Excudates >= 75% granulation** –
1st choice Wound honey with A&D cover with Mepilex border or border dsg change q48
2nd choice Wound gel with Alginate with Mepilex border or border dsg change q 48

3rd choice Iodosorb to wound bed cover with alginate and dry dsg change q 48

For **Mild Exudates' >= 75% Slough** –
1st choice Triad cream with A&D cover with border dsg change daily
2nd choice Bactroban and Santyl cover with alginate and border dsg change daily

Moderate or Large Exudates'

For **Moderate/Large Exudates' >= 75% granulation** –
1st choice Wound honey with A&D cover with alginate border dsg change daily

2nd choice Wound gel with Alginate ag with border dsg daily

3rd choice Iodosorb to wound bed cover with alginate and dry dsg change daily

4th choice Arglaes powder or Comfeel powder with alginate and border dsg change daily

For packing can use Mesalt instead of alginate

For **Moderate/Large Exudates' >= 75% Slough** –

1st choice Polysporin powder with Santyl cover with border dsg change daily

2nd choice Bactroban and Santyl cover with alginate and border dsg change daily

3rd choice Iodosorb with Santly to wound bed cover dry dsg change daily

For packing post surgical wounds use Hydrofera Blue

All Excoriations - Use Calmoseptine initially if not resolving add antifungal cream with 1% Hydrocortisone. If looking red use Metrogyl cream to skin and wrap or cover with Unna Boot Strips.

For Painful wounds
Apply Polysporin powder with Santyl to wound bed cover with a piece of xerofoam and dry dsg change daily. Polysporin helps cut down wound odor, xerofoam will decrease the wound pain and santyl will aid with debridement and tissue repair.

I also sometimes use Metrogel with Santyl to wound bed if there is fecal incontinence like Clostridium difficale since this might be contaminating the wound. For external packing we can use either alginate, mesalt, unna boot strips whichever is in formulary in our facility.

Other options are to use Bactroban or Iodosorb with Santyl to wound bed.

Powders available for wound beds are Arglaes which is a silver powder, multidex powder, gold dust powder from Southwest technologies, comfeel powder from coloplast. Powders with ointment works well if the exudate is large.

For **post surgical wounds Hydrofera blue dressing** with Santyl works well to control wound contamination and debridement.

For wounds that looks like above without slough Silver dressings works very well. Example of silver dressings acticoat, acticoat absorbent, seasorb ag, maxsorb ag, aquacel ag etc.

When using Silver products do not use Santyl simultaneously. Silver will denature the enzymes in Santyl which would become ineffective.

Silver dressings works well with purulent drainage in a wounds or any infected wounds. Silver is expensive dressing and covered only 3 times a week so dressing changes should be kept to q 72 hours. The other option while using silver dressing is to use a small piece of silver dressing on the wound bed and cover the rest of the wound with regular alginate to cut down the cost of wound dressings.

While using silver dressings do not use more than 2 weeks at a time and rotate the products after 2 weeks to accelerate wound healing.

After prolonged usage of the same product the local bacteria creates a biofilm and becomes unresponsive to the antimicrobial products.

Trunk Wounds

Sacral wounds and any pressure wounds if you see the slough with deeper ulcers the wound size, drainage, odor and wound bed determines the need for various dressing products.

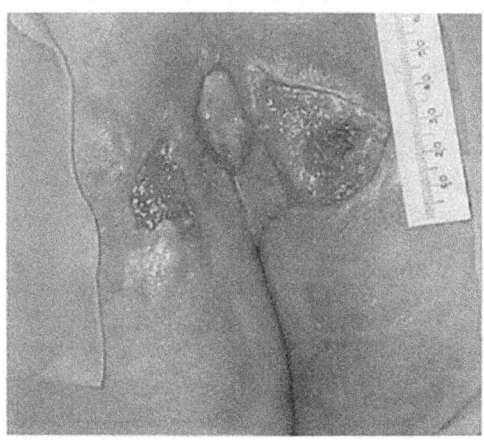

For Painful wounds

Apply Polysporin powder with Santyl to wound bed cover with a piece of xerofoam and dry dsg change daily. Polysporin helps cut down wound odor, xerofoam will decrease the wound pain and santyl will aid with debridement and tissue repair.

I also sometimes use Metrogel with Santyl to wound bed if there is fecal incontinence like Clostridium difficale since this might be contaminating the wound. For external packing we can use either alginate, mesalt, unna boot strips whichever is in formulary in our facility.

Other options are to use Bactroban or Iodosorb with Santyl to wound bed.

Powders available for wound beds are Arglaes which is a silver powder, multidex powder, gold dust powder from Southwest technologies, comfeel powder from coloplast. Powders with ointment works well if the exudate is large.

For post surgical wounds Hydrofera blue dressing with Santyl works well to control wound contamination and debridement.

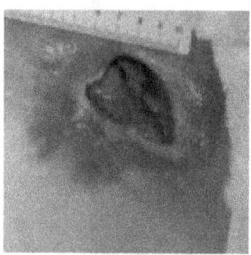

For wounds that looks like above without slough Silver dressings works very well. Example of silver dressings acticoat, acticoat absorbent, seasorb ag, maxsorb ag, aquacel ag etc.

When using Silver products do not use Santyl simultaneously. Silver will denature the enzymes in Santyl which would become ineffective.

Silver dressings works well with purulent drainage in a wounds or any infected wounds. Silver is expensive dressing and covered only 3 times a week so dressing changes should be kept to q 72 hours. The other option while using silver dressing is to use a small piece of silver dressing on the wound bed and cover the rest of the wound with regular alginate to cut down the cost of wound dressings.

While using silver dressings do not use more than 2 weeks at a time and rotate the products after 2 weeks to accelerate wound healing.

After prolonged usage of the same product the local bacteria creates a biofilm and becomes unresponsive to the antimicrobial products.

Extremity Ulcers:

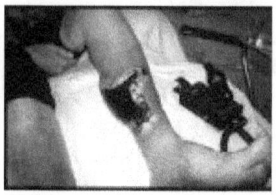

For any extremity injury secondary to trauma if there is evidence of hematoma like above use silver to cover wound bed and apply either Mepilex border dsg or Allvyn gentle with border dsg and change q 72 hours.

I like silver dsg to wound bed and to wrap with unna boot as 12 inch long strips layered with kerlix wrap to help with secondary lymphedema and change this q 72 hours.

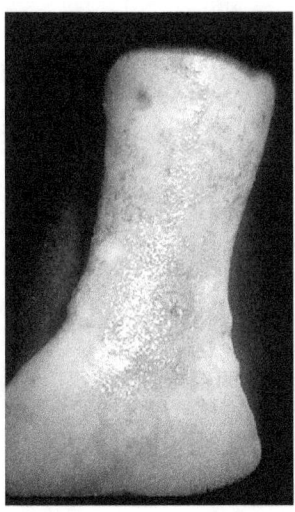

Venous dermatitis use Metrogel with Antifungal cream to skin bid and if there is swelling wrap with unna boot as 12 inch long strips with kerlix change q 72 hours.

If the skin is not red then use Antifungal cream with 1% Hydrocortisone cream bid x 14 days.

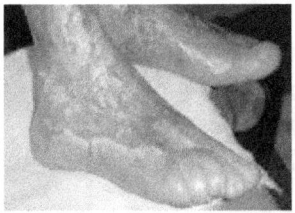

Venous ulcers or combination of venous and arterial ulcers with slough on the wound bed use combination of Polysporin powder with Santyl and xerofoam with unna boot as 12 inch strips as wrap with kerlix dressing to be change every other day.

The other combinations that can be use are Bactroban or Metrogel or Iodosorb with Santyl to wound bed and cover with either alginate, contreet or mesalt dressings.

For painful wounds can add Lidocaine gel to wound bed. For hospice patients can use Morphine gel to wound bed.

Excessive drainage in wounds consider wound vac therapy. Wound vac can be done with Silver foam to help with wound healing. Wound vaccum is usually set up at 125mm hg and dressings are changed 3 times a week.

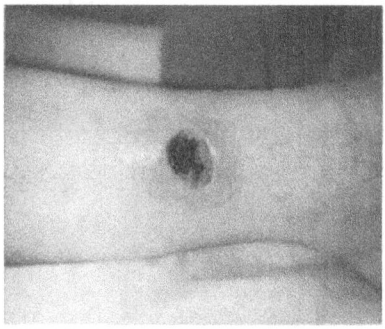

For arterial wounds like above use wound gel with silver dsg and cover with unna boot as 12 inch strip and wrap with kerlix change q 72 hours.

Diabetic ulcers
For diabetic ulcers start with Silver dressing and apply wound gel to wound bed and if possible change dressing daily or every other day.

Dry gangrene:
Just use skin prep to keep the scab dry which is serving as a band-aid provided by the body. Once the healing occurs the scab will lift off and fall eventually.

Kennedy Terminal Ulcer:
If patients ulcer develops suddenly in the presence of deteriorating medical condition think of Kennedy terminal ulcer.

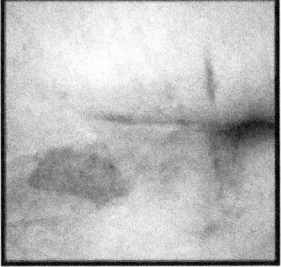

Initially it will look as though there is as though the area has deep tissue injury and in a short while within a day or two can have deep ulcer showing evidence of tissue failure from inside out. The area will feel boggy to touch and the color will look as though the region is not being perfused. Document the development of the lesion and keep monitoring the patient.

Treatment of Kennedy ulcer is similar to the appearance of sacral ulcer if the skin is intact just apply unna boot strip as 4x4 and cover with either Mepilex border or Allvyn gentle dsg and change daily. If there is excessive soilage use border dsg and change daily and prn soilage.

Once the tissue breakdown starts follow the treatment according to the appearance of the wound bed.

Expensive Dressings like BioStep or BioStep Ag Usages

BioStep and BioStep ag is a once in 6 day dressing very useful for blood blisters, non healing diabetic ulcers and ulcers that are clean but not closing for prolonged periods of time.

Use the BioStep product with Mepilex dressing or Allvyn Gentle dressing and change once a week.

TYPES OF DRESSINGS
Not all inclusive

ALGINATES

- AlgiDERM Calcium Alginate Wound Packing
- ALGOSTERIL Alginate Dressing
- Curasorb
- DermaSORB Spiral Hydrocolloid/Alginate Wound Dressing
- KALTOSTAT Wound Dressing
- Sorbsam Topical Wound Dressing

COMPOSITES

- ALLDRESS
- Covaderm Adhesive Wound Dressing
- Covaderm Plus Adhesive Barrier Wound
- Dressing
- EPIGARD
- Lyofoam A
- NU-DERM Foam Island Dressing
- Poly Mem
- Telfa Adhesive Dressing
- Telfa Island Dressing
- Transorb
- Ventex Wound Dressing
- Viasorb Wound with Absorptive Pad

EXUDATE ABSORBERS

- Bard Absorption Dressing
- Bard Absorption Dressing Easy Pack Gel
- Chronic Cure
- Dermanet Wound Contact Layer
- HydraGran Absorbent Dressing
- Multidex Hydrophilic Powder Wound Dressing

FOAMS

- Regular
- Allevyn Cavity Wound Dressing
- Allevyn Hydroiphilic Polyurethan Foam Dressing
- EPI-LOCK Wound Dressing
- HYDRASORB Sterile Dressing
- Lyofoam
- Lyofoam T
- MITRAFLEX, MITRAFLEX PLUS, MITRAFLEX SC

Carbon-Impregnated

- Lyofoam C
- Odor Absorbent Dressing

HYDROCOLLOIDS

Regular

- Comfeel Ulcer Care Dressing
- HydraPad Wound Dressing
- Restore Wound Care Dressing
- Sween-A-Peel
- Tegasorb
- Ultec Hydrocolloid Dressing

Pastes

- Comfeel Paste
- DuoDERM Hydroactive Paste

Granules

- DuoDERM Hydroactive Granules

Powders

- Comfeel Powder
- Arglaes Powder
- Multidex Powder
- Gold Dust Powder

Special

- Actiderm Dermatological Patch
- Comfeel Contour Dressing
- Comfeel Pressure Relief Dressing
- Comfeel Transparent Hydrocolloid Dressing
- DuoDERM CGF
- DuoDERM CGF Border
- DuoDERM Extra Thin
- Orahesive Oral Bandage
- Restore Cx Wound Care Dressing
- Restore Extra Thin Dressing
- TRIAD (amorphous)

HYDROGELS

Amorphous

- Biolex Wound Gel
- Carrasyn Spray Gel Wound Dressing
- DuoDERM Hydroactive GEL
- HYPERGEL
- IntraSite Gel
- NORML GEL
- ROYL-DERM Wound Hydrogel
- WOUN' DRES

Impregnated Gauze

- CarraGauze Sterile 4X4
-

Sheet

- Aquasorb Hydrogel Wound Dressing
- Elasto-Gel
- Elasto-Gel Occlusive Dressing
- NU-GEL Wound Dressing
- Vigilon Primary Wound Dressing

Skin Sealants

- AllKare Protective Barrier Wipe

Hollister Skin Gel

GAUZES

Impregnated

- Adaptic Non-Adhering Dressing
- CarraGauze Strips
- Curity Oil Emulsion Dressing
- Curity Saline Dressing
- Curity Xeroform Dressing
- DERMAGRAN West Dressing (Saline)
- DermAssist Oil Emulsion Dressing
- DermAssist Packing Strips (Iodoform)
- DermAssist Petrolatum Gauze U.S.P.
- DermAssist Wet Dressing (Saline)
- DermAssist Wet Dressing (Water)
- DermAssist Xeroform Petrolatum Gauze
- Mesalt
- Mesalt Ribbon
- Scarlet Red Ointment Dressing
- Vaseline Petrolatum Gauze
- Xeroflo Guaze Dressing
- Xeroform Petrolatum Guaze
- EXU-DRY Wound Dressing
- Sofsorb Wound Dressing
- Telfa Non-Adherent Dressing

Non-Woven

- Excilon All Purpose/Drain Sponge
- NU-GAUZE General-Use Sponge
- Sof-Wick Dressing Sponge
- Versalon

<u>Packing</u>

- Curity Cover Sponges
- Curity Sterile Gauze Pads
- DermAssist Packing Strips (Plain)
- Kerlix 4X4 Sponges
- Kerlix Packing Sponges
- Kerlix Rolls
- Kerlix Super Sponges
- Packing/Debriding
- NU-BREDE Packing and Debridement Sponge

<u>Wrapping/Roll</u>

- Kling Conforming Gauze Bandage
- SOF-BAND Bulky Bandage
- SOF-KLING Conforming Bandage

HYDROGELS (Continued)

<u>Skin Sealants</u>

- Hollister Skin Gel Protective Dressing Wipe
- Prep-Site
- Skin Prep
- Sween Prep

TRANSPARENT FILMS

- ACU-derm
- BIOCLUSIVE Transparent Dressing
- Blister Film Transparent Dressing
- DermAssist Transparent Site Dressing
- HydroDerm Breathable Transparent Dressing
- OpSite
- Polyskin II
- Pro-Clude Transparent Wound Dressing
- Tegaderm
- Tegaderm HP
- UniFlex

WOUND CARE SYSTEM

- DERMAGRAN Spray and DERMAGRAN Ointment

CLEANSERS

- Biolex Wound Cleanser
- CARA-KLENZ Dermal Wound Cleanser
- Clinical Care
- Constant-Clens Dermal Wound Cleanser
- MICRO-KLENZ
- PURI-CLENS
- ROYL-DERM Wound Cleanser
- SAF-Clens Chronic Wound Cleanser
- SEA-CLENS
- Shur-Clens
- ULTRA-KLENZ

POUCHES

- Hollister Wound Drainage Collector

OTHER

Biosynthetics

- Inerpan Temporary Wound Dressing
- Silon Wound Dressing (Non-Adherent Film)

Debriding Agents

- Collagenase Santyl

Topical Sprays

- Granulex Spray

TAPES

- Hypafix Dressing Retention Sheet
- Hy-Tape
- Mefix Tape
- Micropore Tape

Specialty Dressings:
    HydroFera Blue
    BioStep and BioStep Ag
    Contreet
    Biatin Dressings

Wound Care Resources
1. WoundSource.com

Summary

In summary *wound healing requires an approach that is:*

Patient centered: It is always wise to remember that we are dealing with a person who happens to have a chronic wound. We can develop a wonderful management plan but if we do not have patient buy-in the plan is doomed to failure.

Holistic: Best practice requires the assessment of the whole patient, not just the "hole in the patient". All possible contributing factors must be explored.

Interdisciplinary: Wound care is a complex business requiring the skills of many disciplines. Skilled nurses, physiotherapists, occupational therapists, dietitians and physicians both generalists and specialists (dermatologists, plastic surgeons and vascular surgeons depending on need) are central members of the team. In addition in some settings social work involvement may be important.

Evidence based: In today's healthcare environment treatment must be based on best available evidence and be cost effective.

References:
1Kerstein MD: The scientific basis of healing. *Adv Wound Care* 1997; 10(3):30-36
2Kane D: Chronic wound healing and chronic wound management, in Krasner D, Rodeheaver GT, Sibbald RG. (eds): *Chronic Wound Care: A Clinical Source Book for Healthcare Professionals, Third Edition.* Wayne, PA, Health Management Publications, 2001,pp 7-17.
3MacLeod J (ed): *Davidson's Principles and Practice of Medicine, Thirteenth Edition.* Edinburgh UK, 1981, pp 590-592
4Wahl LM, Wahl SM: Inflammation, in Cohen IK, Diegelman RF, Lindblad WJ (eds): *Wound Healing: Biochemical and Clinical Aspects.* Philadelphia, PA, W.B. Saunders, 1992, pp 40- 62
5Kerstein MD: Introduction: moist wound healing. *American Journal of Surgery* 1994; 167(1A Suppl): 1S-6S
6Lazarus G, Cooper D, Knighton D, Margolis D, Pecoraro R, Rodeheaver G, Robson. Definitions and guidelines for assessment of wounds and evaluation of healing. *Archives of Dermatology* 1994;130:489-493
7Winter GD: Formation of scab and rate of epithelialization of superficial wounds in the skin of the young domestic pig. *Nature* 1962;193:293-294
8Knighton DR, Silver JA, Hunt TK. Regulation of wound-healing angiogenesis: effect of oxygen gradients and inspired oxygen concentration. *Surgery* 1981;90:262-270
9Baxter CR. Immunologic reactions in chronic wounds. *American Journal of Surgery* 1994;167(1A Suppl):12S-14S

10Haimowitz JE, Margolis DJ: Moist wound healing, in Krasner D, Kane D (eds): *Chronic Wound Care: A Clinical Source Book for Healthcare Professionals.* Wayne, PA, Health Management Publications, 1997, 49-55

11Mertz PM, Marshall DA, Eaglestein WH. Occlusive dressings to prevent bacterial invasion and wound infection. *J Am Acad Dermatol* 1985;12:662-668

12Hutchinson JJ, McGuckin M. Occlusive dressings: A microbiologic and clinical review. *Am J Infect Control* 1990;18:257-268

13Turner TD. Hospital usage of absorbent dressings. *Pharma J* 1979;222:421-426

What is Laser Therapy?

Laser Therapy is the application of red and near infra-red light over injuries or lesions to improve wound / soft tissue healing and give relief for both acute and chronic pain. It is now officially referred to as (Low Level Laser Therapy) LLLT. The term "LASER" is an acronym for *Light Amplification by Stimulated Emission of Radiation*.

Laser therapy is the use of monochromatic light emission from a low intensity laser diode (250 Milli watts or less) or an array of high intensity super luminous diodes (providing total optical power up to the 2000 Milli watt range). Conditions treated include Musculoskeletal problems, the arthritis, sports injuries and dermatological conditions. The light source is placed in contact with the skin allowing the photon energy to penetrate tissue, where it interacts with various intracellular bio molecules resulting in the normalization of cellular components. This also enhances the body's natural healing processes.

Laser Therapy is used to:

- **Increase the speed, quality and tensile strength of tissue repair**
- **Give pain relief**
- **Resolve inflammation**
- **An alternative to needles for acupuncture**
-

The red and near infrared light (600nm-1000nm) can be produced by laser or high intensity LED.

The intensity of LLLT lasers is not high like a surgical laser*. There is no heating effect.

The effect is photochemical (like photosynthesis in plants)

Red light aids the production of ATP thereby providing the cell with more energy which in turn means the cell is in optimum condition to play its part in a natural healing process.

*LLLT devices are typically delivering 5mW -1000mW (0.2 -> 1.0 Watts).

How long are the treatments?

Treatments can vary in time from seconds to minutes depending on the condition. Research studies show that there may be a dose dependent response, so it may be more effective to treat at lower doses at multiple intervals then to treat a single time with a high dose.

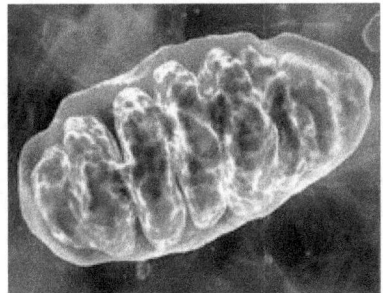

* Pain Relief (muscles, joints, nerves)
* Whiplash
* Plantar fascitis
* Wound Healing
* Trauma
* Arthritis
* Migraine headaches
* Lower back pain
* Repetitive Stress Injuries (RSI)
* Carpal Tunnel Syndrome (CTS)
* Tendonitis
* Fibromyalgia / Myofascial Pain
* Sprains and strains

* Post-operative pain
* Post-operative wounds
* Knee, foot, ankle pain
* Tennis Elbow
* Golfer's Elbow
* TMJ
* Soft tissue injuries
* Swelling
* Burns
* Pressure sores
* Herpes simplex
* Acne
* Rotator Cuff Injury

How does LLLT work?

Like photosynthesis - the correct wavelengths and power of light at certain intensities for an appropriate period of time can increase ATP production and cell membrane perturbation could lead to permeability changes and second messenger activity resulting in functional changes such as increased syntheses increased secretion and motility changes.

Red and near infrared light seem to be the most ideal wavelengths.

Red light and near infrared light acts on the **Mitochondria** and at the **cell membrane**. In in-vitro and animal LLLT wound healing studies comparing wavelengths, red consistently is more effective. Shorter wavelengths are not as good and are more expensive to produce and have poor penetration; overall, they are a poor choice.

Near infrared light, while not quite as good, do penetrate better than the red wavelengths and are available in higher powers and at low prices. According to live in-vivo experiments at Uniformed Services University Bethesda Maryland (a US military research centre) 810nm is the best penetrating wavelength. It also happens to work well in LLLT nerve regeneration studies they are doing.

Clinical Effects of LLLT

An appropriate dose of light can improve speed and quality of acute and chronic wound healing, soft tissue healing, pain relief improve the immune system and nerve regeneration. Applications with good RCT evidence include Venous Ulcers, Diabetic Ulcers, Osteoarthritis, tendonitis, Post Herpetic Neuralgia (PHN, shingles) & postoperative pain.

To paraphrase NASA research:

"Low-energy photon irradiation by light in the far-red to near-IR spectral range with low-energy (LLLT) lasers or LED arrays has been found to modulate various biological processes in cell culture and animal models. This phenomenon of photobiomodulation has been applied clinically in the treatment of soft tissue injuries and the acceleration of wound healing.

The mechanism of photobiomodulation by red to near-IR light at the cellular level has been ascribed to the activation of mitochondrial respiratory chain components, resulting in initiation of a signaling cascade that promotes cellular proliferation and cytoprotection."

"A growing body of evidence suggests that **cytochrome oxidase** is a key photoacceptor of light in the far-red to near-IR spectral range. Cytochrome oxidase is an integral membrane protein that contains four redox active metal centers and has a strong absorbance in the far-red to near-IR spectral range detectable in vivo by near-IR spectroscopy."

"Moreover, 660–680 nm of irradiation has been shown to increase electron transfer in purified cytochrome oxidase, increase mitochondrial respiration and ATP synthesis in isolated mitochondria, and up-regulate cytochrome oxidase activity in cultured neuronal cells."

"LED photostimulation induces a cascade of signaling events initiated by the initial absorption of light by cytochrome oxidase. These signaling events may include the activation of immediate early genes, transcription factors, cytochrome oxidase subunit gene expression, and a host of other enzymes and pathways related to increased oxidative metabolism."

"In addition to increased oxidative metabolism, red to near-IR light stimulation of mitochondrial electron transfer is known to increase the generation of reactive oxygen species. These mitochondrially generated reactive oxygen species may function as signaling molecules to provide communication between mitochondria and the cytosol and nucleus."

HOW DOES IT WORK?

The effects of low energy, Red and Infra-Red light are photo-chemical (not thermal). It triggers normal cellular function.

PHOTONS

ABSORBED IN CYTOCHROMES & PORPHYRINS
WITHIN THE MITOCHONDRIA
AND AT THE CELL MEMBRANE
(Visible red light absorbed within mitochondria)
(Infra-red light at the cell membrane)

↓

SINGLET OXYGEN PRODUCTION
(Rate limiting mechs operate to prevent
excess singlet oxygen formation)

FORMATION OF PROTON GRADIENTS ACROSS CELL
MEMBRANE AND ACROSS MEMBRANE OF MITOCHONDRIA

↓

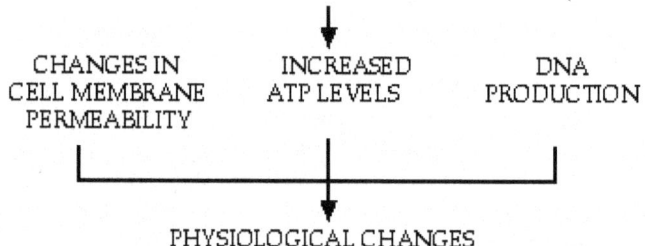

CHANGES IN CELL MEMBRANE PERMEABILITY	INCREASED ATP LEVELS	DNA PRODUCTION

PHYSIOLOGICAL CHANGES

Laser Products: http://www.LightRelief.com or
http://www.MyClaro.com
http://www.coldlasertherapies.com

FDA approved Lymphedema laser:
http://www.Riancorp.com

Listing of current articles and research projects -
http://www.ThorLaser.com or
http://www.BioFlexLaser.com

Meditation with Laser:
http://www.ScalarWaveLasers.com

Principles of Laser Therapy

Low level laser therapy aims to biostimulate. Because of the low power nature of low level lasers, the effects are biochemical and not thermal and cannot cause heating and thereby damage to living tissue. Four distinct effects are known to occur when using low level laser therapy:

A) Growth factor response within cells and tissue as a result of increased ATP and protein synthesis; Improved cell proliferation; Change in cell membrane permeability to calcium up-take.

B) Pain relief as a result of increased endorphin release; Increased serotonin; Suppression of nociceptor action.

C) Strengthening the immune system response via increasing levels of lymphocyte activity and through a newly researched mechanism termed photomodulation of blood.

D) Acupuncture point stimulation.

Parameters for Use

Three parameters are important for clinicians to achieve the best possible therapeutic effects when using low level laser.

A) Selection of the correct beneficial Wavelength

B) The use of the correct Power levels

C) The consistent application of the necessary amount of Energy measured in Joules. Another factor which has to be considered in laser therapy is Pulsing Frequency.

Wavelength

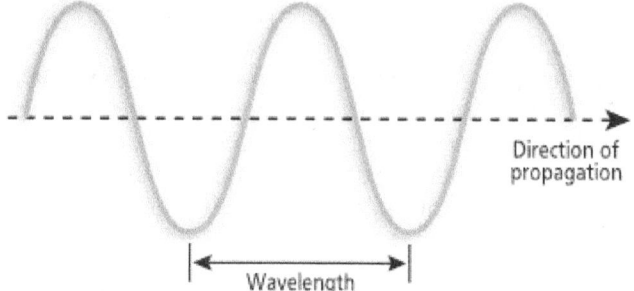

Direction of propagation

Wavelength

Wavelength is measured in nanometers
It is set in the probe and cannot be changed
The 3B lasers are produced with wavelengths between 600 and 1000nm

Red Light Laser Probes with Wavelengths between 620 and 675nm. Readily absorbed by the mitochondria and therefore potentially stimulatory Excellent source of stimulation of a range of growth factors. Red Light does not penetrate very effectively below the skin surface and into the tissue below

Red light is the best for wound healing or superficial conditions but is not the most effective way to treat deeper injury

Infrared (Invisible) Laser Probes with Wavelengths between 780 and 950nm. Absorbed through the cell walls (acting differently between cells) and therefore cell response is more wavelength specific in the infrared range, responding differently to different wavelengths.

More penetrative through the tissue, especially the 780-830nm range, therefore this range is selected for treatment through intact skin and pain relief

Cluster probes with a range of wavelengths offer the best of both worlds to a clinician

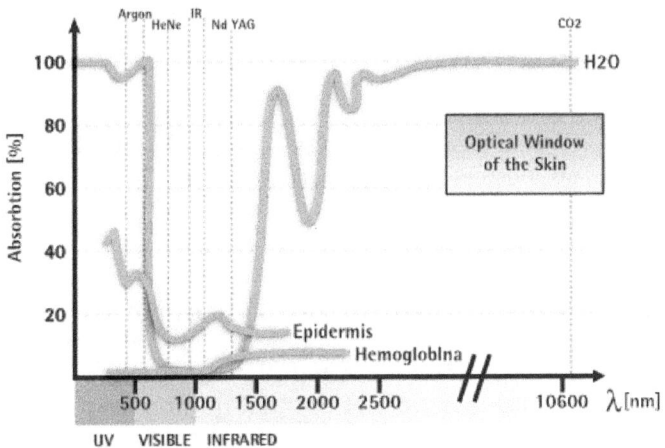

How to Calculate the Necessary Energy

Power (Watts) and Energy (Joules)

Energy = Power × Time

Power (watts or milliwatts) is the strength of the probe and is set at the factory by Omega. Clinicians need to know the power of the probe they are choosing to use and also have the probes regularly checked to ensure that they are running to full power. This check can be quite simply carried out by a special measuring device in the Omega systems.

The simple formulae as outlined above shows the calculation necessary to establish the amount of energy which we as clinicians are providing in each treatment.

The relationship is Energy (as measured in Joules) = Power × Time. Omega provides detailed literature on the most commonly calculated procedures.

To create a response in the growth factor cells or to provide pain relief it is necessary to make sure that we are delivering sufficient energy.

Simply, to increase the energy being delivered by the probe you increase the time of treatment.

Pulsing Rate

Pulsing rate equals the number of times per second light is emitted. Research has demonstrated there are pulsing rate specific effects with low level laser therapy. Concise details of optimal pulsing frequencies and energy density requirements are supplied with our laser equipment

WALT

World Association of Laser Therapy Recommendations:

http://www.walt.nu/dosage-recommendations.html

Recommended anti-inflammatory dosage for Low Level Laser Therapy

Laser classes 3 or 3B, 904 nm GaAs Lasers (Peak pulse output more than 1 Watt)

Energy dose delivered to the skin over the target tendon or synovia

Diagnoses

Tendinopathies	Points or cm2	Joules 904nm	Notes
Carpal-tunnel	2-3	4	Minimum 2 Joules per point
Lateral epicondylitis	1-2	1	Maximum 100mW/cm2
Biceps humeri cap.long.	1-2	2	
Supraspinatus	2-3	3	Minimum 2 Joules per point
Infraspinatus	2-3	3	Minimum 2 Joules per point
Trochanter major	2-3	2	
Patellartendon	2-3	2	
Tract. Iliotibialis	2-3	2	Maximum 100mW/cm2
Achilles tendon	2-3	2	Maximum 100mW/cm2
Plantar fasciitis	2-3	3	Minimum 2 Joules per point

Arthritis	Points or cm2	Joules 904nm	
Finger PIP or MCP	1-2	2	
Wrist	2-3	3	
Humeroradial joint	1-2	2	
Elbow	2-3	3	
Glenohumeral joint	2-3	6	Minimum 2 Joules per point
Acromioclavicular	1-2	2	
Temporomandibular	1-2	2	
Cervical spine	2-3	6	Minimum 2 Joules per point
Lumbar spine	2-3	10	Minimum 4 Joules per point
Hip	2-3	10	Minimum 4 Joules per point
Knee anteromedial	2-4	6	Minmum 2 Joules per point
Ankle	2-4	6	

Daily treatment for 2 weeks or treatment every other day for 3-4 weeks is recommended. Irradtiation should cover most of the pathological tissue in the tendon/synovia.

Tendons
Start with energy dose in table, then reduce by 30% when inflammation is under control. (Does not apply for carpal tunnel tenosynovitis)

Therapeutic windows range from typically +/- 50% of given values

Recommended doses are based on ultrasonographic measurements of depths from skin surface and typical volume of pathological tissue and estimated optical penetration for the different laser types in Caucasians

Disclaimer: The list may be subject to change at any time when more research trials are being published. Meridian Medicine Applications: Laser Therapy is not responsible for the application of laser therapy in patients, which should be performed at the therapist/doctor`s discretion and responsibility.

Laser Treatment for Wound care

While using laser for wounds the treatment should be done at the wound edges to stimulate the wound edges. Direct application of laser on wounds could cauterize the tissue and would cause more tissue pain. If there is pain while applying laser remove the laser and start again 2 cm away from the area of pain lateral to the wound edges.

Wound edges is to be stimulated at a 2 cm interval circumferentially.

If there is increased discomfort while applying laser this is a sign of lymph node inflammation and stop the procedure immediately.

Types of laser for wounds can either be 600nm lasers or LED lights.

Laser Treatment for Lymphedema

FDA approved Lymphedema laser is distributed by Riancorp and is a 904 nm wavelength probe laser. The probe is supposed to work at the level of upto 2cm from the skin.

For Lymphedema start from the proximal area to the heart and go distal towards extremity.

Stimulate the supraclavicular nodes, jugular nodes, axillary nodes, thymus, abdominal nodes then follow up with the extremity you need to work on.

To stimulate use the probe laser at the 2 cm distance from the each point and hold the laser for 30 – 45 seconds. Over exposure could cause lymphangitis and be very prudent when you are working with the lymph nodes.

For extremity that is swollen use the area and stimulate as much area as possible. If there is increased discomfort immediately stop the procedure to prevent tissue damage.

Some Over the counter Laser Resources include

1. www.biohorizonmedical.com
2. Bioflexlaser.com
3. www.laserhealthsystems.com
4. Lazershop.uk.co
5. DragonLasers.com
6. www.thorlaser.com
7. www.scalarwavelasers.com